Free Flig

Celebrating Your Right Brain

Barbara Meister Vitale

Foreword by Bob Samples

JALMAR PRESS
Torrance, California

Library of Congress Cataloging-In-Publication Data
Vitale, Barbara Meister
Free Flight.
Bibliography: p. 109
1. Cerebral dominance. 2. Brain—Localization of functions.
3. Left and right (Psychology) I. Title.
QP385.5.V58 1986   152.3'35   CIP 85-30018
ISBN 0-915190-44-3

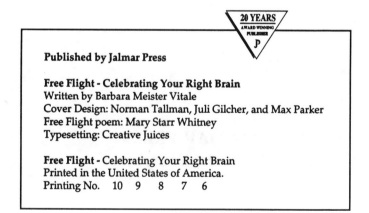

Published by Jalmar Press

Free Flight - Celebrating Your Right Brain
Written by Barbara Meister Vitale
Cover Design: Norman Tallman, Juli Gilcher, and Max Parker
Free Flight poem: Mary Starr Whitney
Typesetting: Creative Juices

Free Flight - Celebrating Your Right Brain
Printed in the United States of America.
Printing No.   10   9   8   7   6

To
God — who made this
book possible
and
Louis Vitale — who helped God
out a lot

Contents

Acknowledgements

I celebrate and thank the many friends and teachers who helped me fly! Special thanks go to:

Bob Samples — for honoring the mother in me.

Suzanne Mikesell — my editor and friend, for being able to get inside my head.

Norman Tallman — for sharing the Lakota concept of the eagle's energy in the cover design.

Mary Starr — for the inspiration of her poetry.

Hugh Prather — for his poetry.

Lois Sanford — for always being there and understanding friendship.

Leo Buscaglia — for encouragement.

Juli Gilcher — for taking me beyond motherhood to friendship.

Rick Meister — for forcing me to understand his right brainedness.

Bob Gast — for showing me what reading really is.

Rick Sutter — for being the most courageous person I know.

Joan Werling — for proving that she can make it.

David Schriff — for learning to succeed as himself.

Susan Hicks — for being special and beautiful.

Michael Wolf — for using his brain to tap the universe.

David Walker — for teaching me what special is.

Michael Bohn — for seeing angels.

Andrea and Steve Swell — for just being beautiful.

Ami Bohn — for drawing flying unicorns.

Janet Lovelady — for helping me to learn patience.

Bradley L. Winch — for having faith in education.

Barbara Meister Vitale
Delray Beach, Florida

Foreword

Ideas are hard to come by. And good ideas are even harder to find. The notion of the differentiated functions of the cerebral cortex is, I suggest, one of those good ideas of this generation. It is an idea supported by medical research, psychology, biofeedback, history, literature, systems theory, and common sense. Not that it is without its detractors and critics. They abound. But for the most part, criticism is characterized by professional territorialism and carping about details.

The lateralization of the human cortex has provided us with the condition called "left and right brain." In most people, the left cortex is specialized to process codes, both written and spoken, which follow specified rules of logic, form, and order. Most often these are the functions that provide symbolic and abstract forms of discourse and expression. The right hemisphere, in most, provides a more diffuse and generalized set of support capabilities to thinking and living. These include ways of knowing that are more synthetic, more affective. They provide a cohesiveness to everyday life and its breadth of experience.

Another, more prosaic, way of saying this would be to say that the left brain attends more to the rules

ix

of conduct and the formalism of society. The right is more attendant to the informal, personal, and emergent ways of knowing that are more related to our natural design. Because of this differentiated commitment in our brain-mind system we are predisposed to a natural dualism. Some prefer the metaphors of war and conflict between the hemispheres to describe the goings on in our heads. Heroic as this may be, I prefer descriptions of complementarity and unity.

In this book, Barbara Vitale takes the reader on an excursion through the richness of her own experiences and observations regarding the duality of mind. With personal courage and honesty she guides us through poignant descriptions of her journey and offers advice on our own.

I feel this book is a viable expression of how the information about the left and right brain can be used as a metaphor for everyday life, including the dualities encountered there. Do not, however, treat this as a medical text or a formal compendium of research on cortical function. Instead, read as one invited into and embraced by Barbara Vitale's life. This book is born of the human soul.

Bob Samples
The Metaphoric Mind

Introduction

FREE FLIGHT shares my discovery of my right-brainedness and how I learned to celebrate its existence. It is a journey, not a destination, a beginning not an end. It is based partly on personal experience and partly on the experiences of others. I invite you to celebrate with me. Regardless of what brain you're in when you read this, remember that our brains are holistic. The two hemispheres work together, creating individual perspectives of our lives. Our brains are as different as our fingerprints, enabling us to be unique. Celebrate that difference! Fly the universe!

"Trust yourself. This is your dream."

Hugh Prather
There Is a Place Where You Are Not Alone

Takeoff
Discovering My Right Brain

Hot and sweaty from playing in the flower garden, I ran across the room to my grandmother's lap. It was a big lap and wonderful to sink into. I called my grandma "Big Mommy." I needed to share! I needed to explain what had just happened to me but how could I explain something I saw, felt, tasted, smelled, and sensed but never heard words to explain? Besides, I was only five and sometimes adults didn't even listen. This time I had to make her listen.

"Big Mommy, there were these kids — two of them — they weren't there but they were real. I know they were. We talked, up in the trees. It was fun. Not really kids, fairies; oh, do you think they will come back?"

Big Mommy just smiled and hugged me. It was OK with her if I believed in something that seemed to make me happy. My fairies only stayed a few years. They went away when I started school. They didn't like school. They were right!

3

School and I started off on the wrong foot. The very first day was a nightmare. I had a beautiful new dress. It was made out of printed feed-sack material, but had lots of secondhand lace. I stood out by the road waiting for the bus. The minutes seemed like lifetimes. My stomach felt like there were tadpoles swimming around in my milk and cereal. Finally I saw the big yellow bus speeding up the road. It didn't look like it was going to slow down, so my "Mommy" waved 'til it stopped. The driver said it wasn't the right bus. I had missed it, but he would drop me at the grade school.

My legs could hardly reach the steps. They hurt as I stretched to pull myself up. I sat in the first seat. I was too scared to move back. The "kids" looked all grown up. The tadpoles felt like they were drowning. We made several stops before the bus-man said, "This here's the grade school." I just sat. "You gotta get out here." He stared at me but he was smiling. I jumped down the steps. There were three. The bus pulled off. I looked around. There were no kids or people, just a lot of steps and a big, big building. I felt disappointed. I was sure school would look like Cinderella's castle. This school just looked like a cardboad box only it was gray and had dirty windows. I climbed the steps, being careful not to step on a crack. I didn't want to break my mother's back.

There were twenty-two and one-half steps — one was broken so I didn't count all of it. I don't remember learning to count, but I knew I could. The top step didn't feel right so I sat on the one below it. It made me cool, but didn't help the loneliness. Big yellow buses began to arrive. Kids got off and walked by me, but nobody spoke. I didn't either. They stopped going up the stairs and all the buses pulled away. I knew I was to go someplace and do something, but I did not know where or what so I just sat! I began to watch the ants moving in and out of the cracks. I could imagine what they were saying to each other. I gave each ant a name. There was George. He seemed the smartest. And Emily — she ran back and forth, getting Tom to follow George.

I took a small piece of peanut butter and jelly sandwich out of my lunch bag and wedged it between two cracks. Boy, were those ants strong. In no time, the bread had disappeared down the crack and I was alone again.

I was watching the clouds make wonderful animals when this man came out of a door in the cardboard box and asked me, "Why aren't you in your room?" What a funny thing to ask. "My room's at home," I said. He began to frown, "Don't you know where you belong?" I thought I belonged inside me, but I didn't think that's what he meant so I just shook my head. "Come on, follow me." I got up and followed him into the cardboard box. We walked down a long hall with lots of closed doors on both sides. I wondered if there were monsters locked behind the doors. We stopped in front of one and the man opened it. I felt really scared. He pulled me in anyway. The room was full of kids, sitting at little brown desks. They all stared. A skinny lady in a bright orange dress said she was my teacher. I didn't like the dress. She said, "You are going to be in my room. What's your name? Have you registered? Where do you live? Why didn't your mother come?" By the time she got through, I had trouble remembering my name.

My desk was at the back of the room. At least it was for a couple of days until I got moved way up front because I talked when the teacher didn't want me to. It seemed that was all the time. I couldn't understand everything she was talking about, so I'd ask another kid what she said. That's when I got yelled at. Boy, could she yell! She kept saying she didn't understand why I wasn't understanding but she wouldn't let me ask. So I just kept not understanding.

My second grade teacher didn't yell as much. I guess she had figured out I was dumb. It was funny. I never felt dumb 'til I started school. That dumb feeling stayed with me for a long time.

On rainy days, my second grade teacher used to let us dance to records in this big room. I loved to dance. I forgot anyone else was even in the room. I'd shut my eyes and pretend I was one of my fairy friends playing in the clouds. The other kids laughed but I didn't care.

Our class was going to give a play for our parents. All the kids were in the play but two of us. I don't think he could read either. The teachers said I could dance to my favorite record on stage. I couldn't believe it. I was so excited!

On the day of the play the teacher broke the record. The teacher said she was sorry, but I knew she really didn't want me to dance or she wouldn't have broken the record. I remember I wanted to die. I didn't go to the play. That's all I remember about second grade.

In third grade we moved. Just before we moved I learned how to cheat. I also knew I couldn't bear having the other children snicker when the papers were passed back and I always got a red "F." I still hate red. I figured if I wrote the words on a piece of paper and put them just inside my desk top, I could break my pencil and pretend to be getting another. I got caught. The teacher made me miss recess to write the words 500 times each. I wrote real slow. By the time we moved, I had finished two words. Today I am a creative speller.

I remember a little more in fourth grade. I remember giving my first book report. I couldn't write it down or read it, but I memorized a story I heard on the radio. The teacher said it was good, but I got an "F" because I didn't turn in the written work.

My grades were beginning to come up. I learned to cheat and not get caught. On rainy days we used to play a game called "Mr. Ree." It had something to do with finding out who murdered whom and where he hid the weapon. I never did figure out how to play, but I always knew who was guilty and where they put the weapon. I always won. Pretty soon the other kids didn't want to let me play.

I remember fifth grade. I learned to read. Mrs. Grosh was a little white-haired teacher with a loud voice. When she got mad at us, she would slam a book on the desk and stare at us. We got very quiet. I don't remember her ever saying anything mean to any of us. The whole class had to make scrapbooks for Geography. When she found out I couldn't read, she asked, "If you could go anyplace in the world, where would you go?" I had seen this wonderful picture of the rain forests and mountains. "I'd really love to see Washington State." "Then we will go together." She helped me send away for pamphlets, find pictures, and put the scrapbook together. Most important, she taught me to read the words under the pictures. Do you have any idea what it feels like to suddenly begin to understand that what you say and think can be written down? That you can explore what others are thinking, that you can travel anyplace you want, that you are no longer ever alone!

I visited Mrs. Grosh several times before her death. Each time, she pulled out her years of grade books and beside each student's name she had written where they lived, who they married, how many children they had, and what they had done. These trips were special for me. I needed to let her know the joy she had given me.

Throughout my junior high and high school years, I still experienced times of confusion. I remember the humiliation when my dress dissolved in Chemistry class. I had mixed up the order of the chemicals. (Sometimes I still do this with recipes.) There I stood in a dress full of holes.

I had many other experiences. I was lucky enough to go to a small high school where competition was limited, at least in numbers. There were forty in our graduating class. I survived academically by developing strategies I later learned were right-hemisphered. I also dated smart boys who helped me through my left-hemispheric homework. I never would have passed Latin if my first love had not helped me translate every night. My self-image stayed intact only because I became involved in cheerleading, drama, and speech.

All my life I have felt different but I never quite understood why. I just knew I didn't think like most other people. Sometimes I knew answers without really knowing where they came from. The year I became a teacher of special children, I began to discover the key to knowing myself. I attended my first conference for the Association for Children and Adults with Learning Disabilities. The conference's main theme had to do with brain research. Since I knew nothing about the subject, I decided to attend all the lectures in this area.

The first one was at 8:00 a.m. For me, that's an unholy hour. I walked in half asleep, hoping to stay that way. It wasn't long before I found myself sitting up straight and leaning forward. Wait a minute! What was the speaker saying? He was talking about a group of people he called right-hemisphered, but it sounded like he was describing me. If what he was saying was true, then I wasn't dumb like I had always felt. I wasn't different. There were others who had thoughts like mine. Could it really be possible! Oh, God, please let it be true!

On that day of discovering the right-hemispheric mode of thinking, I felt my life turn upside down. So many things fell into place. I felt anger at having suffered so much during those early school years, at having wasted so much time feeling different. Mostly, though, I felt joy that now I could understand and appreciate my special abilities. I decided to believe in myself, to celebrate my right-brainedness, and to help others recognize and value their right-brained approaches to life.

I wanted to shout to the world, "God doesn't make mistakes!"

14

"And between your knowledge and your understanding, there is a secret path..."

Kahlil Gibran
The Garden of the Prophet

Flight Coordinates
Brain Anatomy and Specialization

I began my quest by buying out the bookstores. To my surprise, I found a number of books written on hemispheric specialization, modes of consciousness, handedness, and the triune brain.

I couldn't read fast enough. Each book or article helped me become more aware of who I was.

When my brain went on overload, I began to travel. I visited Beth Israel Hospital in Boston, Massachusetts, where Dr. Albert A. Galaburda* conducts research on the brain. As I stood in his laboratory looking at a man's brain, I felt awed. Although it looked like any other body organ prepared for a laboratory study, it differed in one crucial respect. Scientists understand how most other

---

*Dr. Galaburda is Director of Dyslexia Laboratory at Beth Israel Hospital and Assistant Professor of Neurology at Harvard Medical School.

organs work — but they know very little about the brain. Mankind has yet to discover the secrets of the source of intelligent thought. I wanted to know everything! I wanted to understand my own brain and how it worked.

From reading, I discovered that the brain weighs about three pounds and has approximately 10,000,000,000 nerve cells. Brain size varies and has little to do with intelligence. Although Einstein was reported to have a large brain, one of the largest brains belonged to an idiot. Women's brains tend to be smaller than men's, which does not lead to the conclusion that men are smarter.

The brain looks somewhat like soft cheese poured into a walnut-shape mold. The outer surface of this walnut is called the cerebrum. The brain is split lengthwise down the middle, forming a narrow gorge that divides the cerebrum in half. These two halves are called the left and right hemispheres. Deep within the brain they are connected by a bundle of nerve tissues called the corpus callosum. The corpus callosum is the hemispheric integrator. That is, it acts as a telegraph line between the two hemispheres. Without the corpus callosum, integrated thought (whole-brain thinking) would be impossible.

As I read further, I learned that although the two halves of the brain appear identical, researchers have found structural and functional differences between them. Some regions of the brain are organized bilaterally symmetrical while others are organized asymmetrically. Also, different areas of the brain are responsible for specific abilities, functions, and parts of the body. The regions of the brain responsible for specific tasks are located either in one hemisphere or the other. Although language develops in the left hemisphere for the majority of people, it can develop in either hemisphere. The speech and hearing centers are on the left side of the brain just above the ear (Broca's area). The area primarily responsible for sound memory is behind the hearing center. Your ability to discriminate sounds and to use phonics in reading is considered mostly in the left hemisphere. Could this be why I had always had difficulty sounding out words?

The area of the brain that controls body movement runs across the top of the head on both sides of the brain. It is called the motor cortex. Located just behind the motor cortex is the area of the brain that handles sensory information. The sensory areas receive and process information from the skin, bones, joints, and muscles and from the movement of your body through space. This process involves the five senses of tasting, touching, smelling, seeing, and hearing. The sensory areas receive and process information from the inner ear (vestibular system) and internal organs (visceral system). The sensory and motor regions together are sometimes called the haptic areas. These regions are organized so that the left half of the brain controls the right side of the body and the right half controls the left side of the body. Simply stated, your left brain moves your right hand. Both the motor cortex and sensory regions are also specialized so that localized sites can be associated with specific parts of the body. These sites are identical in each hemisphere (bilaterally symmetrical). The motor-sensory areas are especially active when you are moving.

Teachers had always considered me hyperactive, which meant I moved all the time. For awhile, I thought I had something wrong with me that made me move. Now I understood. People who move a lot are taking in information through the part of the brain that controls sensory input and movement, allowing them to learn. They learn by doing. Education now calls these movers "haptic learners" because they process information primarily through the motor and sensory cortex of the brain.

The area of your brain primarily responsible for vision, the occipital area, is at the back of the head just above the end of the spine (the brain stem). The occipital area is identical in both hemispheres. It is part of the visual cortex. Different brain cells in this area register messages from different parts of the retina and process the information in terms of size, shape, color, position, and distance. The vision area processes color. Did this begin to explain why certain colors seemed to affect my learning?

The information from the retina of the eye is carried to the occipital area by the optic nerves through a complex process. The vision field of each eye is divided in the middle by the optic nerve. Near the back of the eyes, the optic nerves meet (optic chiasma). The nerve connection from the left side of the right eye and the right side of the left eye cross. This enables each hemisphere to receive information from both eyes. The left side of both eyes sees those objects located to the right of your nose and projects them to the left side of your retina. These images are in turn relayed to your left hemisphere. The right side of both eyes sees everything to the left of your nose and sends information to the right hemisphere. In addition, one eye will tend to be dominant, meaning it initially takes in more information from the environment. The dominant eye thus sends a greater volume of information to its controlling hemisphere. The right eye sends a greater percentage of the information it receives from both visual fields to the left hemisphere than it does to the right.

I wasn't sure at this point what right-brained was, although I felt I was right-brained. My eye doctor had told me I was right-eyed which meant a greater percentage of visual information was going to my left brain. Was this confusing my brain?

The function of the ears is to transfer sound to the brain's "hearing center" (temporal lobe) where it can be interpreted. This is done by collecting the sound waves and moving them to the ear drum. The drum vibrates, setting up a chain reaction that sends electrochemical signals to the brain stem. Here, many of the nerves cross, sending signals to both hemispheres. Although sounds received in each ear go to both sides of the brain, information given to the right ear is sent first to the left hemisphere and then to the right. Each of us has a dominant ear. My left ear is dominant. I have found that turning my left ear towards any sound or conversation helps me to better understand what I hear.

There are different theories explaining hemispheric specialization. While researchers disagree as to the age that specialization of brain functions occurs, they do not dispute the fact that specialization does occur. They also agree that most people have a dominant hemisphere. As the brain specializes, the left hemisphere becomes better at certain skills and the right becomes better at a different set of skills. Between five years of age and puberty, humans develop a dominant hemisphere. Although we have two hemispheres, we tend to use one — the dominant one — more than the other. When you talk, write a letter, or drive a car, you use both sides of your brain. In fact, you use your entire brain for everything you do. Nevertheless, the dominant side of your brain processes information first or is the primary center for processing certain types of information. The dominant hemisphere is the one that activates first and handles the greatest percentage of responsibility for a specific task within a whole brain process.

We must not over-emphasize the significance of the dominant hemisphere. Being right-brained does not mean you do not use your left hemisphere. For many people there is a balance between hemispheres, with each taking control of the tasks it is best at handling. Nevertheless, research has indicated that your dominant hemisphere may determine the skills in which you excel, the way you approach life, and even the way you handle stress. Scientists are discovering more about the brain every day.

All of this physiological information was great. I understood how my brain is specialized but I didn't understand how this specialization and dominance affected me personally. Back to the books. This time I explored the studies involving the skills and behavioral aspects of hemispheric dominance. These studies indicated that differences exist between people who have a strong dominant left hemisphere and those who are called right-brained. I found that certain academic skills appear to be assigned to the left or right hemisphere. Left-brained people are usually good at writing — that is, they write well and can express their thoughts on paper. They are good at interpreting symbols such as letters and words. They are good at language. They are good at labeling things and have good vocabularies.

I began not only to gain insight into myself but also to understand my husband. He appeared to be left hemisphered. He has an extensive vocabulary and writes the most beautiful business letters and memos. Recently I found myself on a job that required me to write a number of memos. I had not yet learned how to use my right brain to do left-brained skills. I consequently found myself agonizing over one memo for hours, unable to put my thoughts into sequential order. Finally, I had my husband write for me. My boss never knew the difference and thought I was wonderful.

I remember in school wondering how the other students could read a science chapter and remember what was in it to answer in class. At the time, I didn't realize these were left-brained strengths. I would read the chapter, get to the end and not have the foggiest idea what I had read. If I was lucky enough to remember any of it, I would have to keep saying those answers over and over in my head until the teacher asked the question that fit my answer. She was good at asking the wrong question. Regardless of what she asked, she got the answer that was in my head.

Left-brained people usually do well in reading as it is taught in our schools. They are good in phonics. I wasn't. I didn't learn the phonic sounds until I had to learn how to teach them to my first grade readers. I still can't use them to decode words. I memorize words. Words like obsequious, proselytize, and reconnoiter might as well be written in Greek.

For most people, the auditory center is in the left brain. Those with a dominant left hemisphere have good listening skills and can interpret what is being said. They are good at following directions. It suddenly dawned on me why my husband loves to explain things to me and why I don't understand. I have tried to explain to him that the words go in but they just seem to float around. I hear them but they don't make any sense. If he physically shows me or draws me a picture, I have no difficulty understanding.

In school, we are graded on reading, writing, language, phonics, reciting, listening and taking tests, all left-brained strengths. If you are left-brained dominant, you probably were an "A" or "B" student who survived the educational system intact. If you are right-brained like I am, you probably had your self-image somewhat damaged.

Thank God I didn't stop here! As I read, a whole new set of skills began to unfold for me. A set of skills I was good at. I had already discovered why I was a "mover," unable to sit still without moving some part of my body. All my report cards said, "doesn't sit still," "plays at desk," and "talks too much." I was good at haptic awareness. Being haptic means more than being "hyper." It means you have a need to move. You are a toucher. You must feel people. You must take their hands in order to get to know them. You are a hugger. You may go into sports, especially an aggressive sport where you can have body contact. Although the motor region of the brain is in both hemispheres, the ability to make judgments based on the relationships of our bodies to space, so needed in sports, basically is centered in the right hemisphere. In high school, boys who are good at haptic awareness and spatial relationships are on football teams but they often are failing English. These same students may be good in Geometry but failing Algebra.

Being good at spatial relationships may help right-brained people to become good artists, mechanics, architects, or interior decorators. Right-brained people are good at dealing with shapes and patterns. They not only like to decorate, they are good at it. They are good at visualizing in their heads total rooms and environments, as well as distinguishing between different colors and hues. Computer programmers tend to have strong right-brained abilities. They visualize the whole program before they ever start. Right-brained people are usually good at math. They may excel at math, especially math computations, but have difficulties in reading and language.

Although singing and music are primarily in the right brain, the ability to read and compose music is left-brained. Our great composers could be considered among the whole-brained. Many right-brained people enjoy singing. Since the auditory center is in the left brain, some of us have difficulty staying in tune. But we sing anyway. We'll sing under our breath; sing during the sermon; or we hum during a class and of course in the bathtub. Some of us need to hear music in order to learn or need to have noise around us all the time.

Right-brained people are good at art expression. They doodle on everything at any time. They also are talented at such skills as woodcarving, sculpture, painting, or illustrating. Creative art includes any art form that does not follow an exact pattern, for example, flower arranging, creating clothes, or designing a house.

Right-brained people are creative in all kinds of ways. They are good at visualizing, which means they daydream a lot. Right-brained people can enjoy themselves no matter where they are because they have this whole world they have created inside their heads. If they get bored, they start turning on the pictures. They go someplace fishing or to the nearest beach and thoroughly enjoy themselves. They can look right at the boss, the teacher, or even the minister and appear to be attentive yet not even be there at all. Right-brained people tend to be emotional. They tend to be sensitive to other people because they are able to recognize and process other people's body language. Although sensitivity is a gift when used to evaluate a person or situation, it doesn't seem like a gift when the information received is critical. Many times I have experienced rejection or hurt from a person who outwardly appears to like me. People can say anything they choose, but I always know what they feel.

Right-brained people are color sensitive. I get up in the morning and get dressed, then I'll say, "This color doesn't feel right. This is not the way I feel," and off comes everything. I can't tell what color feels right until I get it on. So, by the time I am ready to leave in the morning, the bed is covered with clothes! Sometimes I get them hung up and sometimes I don't, but when I walk out the door I'm wearing a color that makes me feel the way I need to feel.

As I began to explore other aspects of the brain, I discovered the psychological theories beginning to surface. The first book I read was Ornstein's *The Psychology of Consciousness*. Ornstein proposed that each hemisphere of the brain specializes in a different mode of consciousness or way of thinking. That is, left- and right-brained people approach life in a different way. This can be stressful, especially if you're right-brained and married to a left-brained person. Left-brained people handle life in a linear fashion, which means they start at the beginning and work until they end up with logical results. The problem with right-brained people is they have to know where they are going before they can get started. Consequently, if no one shows them where they are to end up, they may never get started. They work and think whole-to-part, holistically. When I am learning about anything new, I have to read several books before any of it begins to make sense. When I wrote *Unicorns Are Real*, I taped six months of lectures, jotted notes down on everything from napkins to airline tickets, and still couldn't get started. Finally, I wrote the table of contents and the conclusion. Then I knew where I wanted to go with the book. I got four boxes, labeled the boxes by chapter, and threw the notes in the right box.

The left brain thinks in symbols. The left-brained person can understand a map; the right-brain person has to drive. Right-brained people need to do it to understand it. They need to touch it, feel it; they need to have the real thing in their hands or get as close to it as they can. If you start talking abstracts to a right-brained person, you usually will lose them. A noun to the right-brained person is not a person, place, or thing, but, rather, something they can see, touch, taste, feel, or smell.

When I took driver's education, I was taught that you do not slam on the brakes when you hit ice. I remember getting the answer right on the written test. I also remember the day I hit ice and ended up plastered against the guard rail. I learned the concept when I experienced it.

Left-brained people are sequential. They get up in the morning and start making a mental list of activities in the order they intend to do them. They may think, " I have to get some gas and I'm going to the bank and I have an appointment at nine o'clock and another at ten o'clock and I have a luncheon date, etc." I get up and say: "What do I have to do today? I have to pick up a roast for dinner and I must get the mail and I have to go to work today and if I don't get some gas on the way to work, I'll run out and I have to call my husband at three o'clock to remind him to bring milk home for the party." I keep randomly thinking of everything I have to do. I do not move through the tasks sequentially.

I can't clean the house in sequence either. When I clean house, I pull everything out. A little bit from this room, a little bit from that room and a little bit from who knows where. When my husband walks in, he walks right back out again. And I say to him, "If you don't like the way I clean the house, you clean it. I don't like to do it anyway." I am not a sequentially organized person. I don't think sequentially. I can't even write sequentially. My head is jumping all over. Most of the time, my thoughts are going faster than my pencil.

The left brain is logical. My husband is logical. If there is a logical reason, he will find it. If it is not logical, don't try to explain it to him. Right-brained people don't function that way. They function with gut-level feelings. My husband will say, "Well, how do you know?" I don't know how I know — I just know. I'll never forget when he brought home two stocks and said to me, "I'm going to invest in one of these." I said, "Do not invest in that one." He said, "Give me a logical reason." I said, "I can't — just don't invest in it." He invested anyway and lost money. Right-brained people do not always know where their intuition comes from. Scientists are just now beginning to understand the right-brained intuitive capabilities. Everyone is intuitive, but if you are right-brained, you seem to have more intuition. You cannot put into words where these ideas come from. If you have them, listen to them.

Left-brained people are verbal — they are good at describing incidents, have large vocabularies, and talk in complete sentences. Right-brained people are non-verbal, which means they talk a lot but have difficulty saying what they are thinking. They talk in phrases. They talk with their hands. They have trouble putting labels on things. I'll say, "Louis, bring me that, you know, that thing that's about this long — that thing you use to pick the crumbs up — you know, it's sitting in there on the shelf — you know what I am talking about, it's brown and white and it's on the second shelf." He'll say, "Barbara, what are you talking about?" I'll say, "You know, that little vacuum cleaner." He'll say, "Oh you mean the Dustbuster." "Yeah, that's what I want." Sometimes I can't get the words out. If you look at the research, you'll see that the right brain can talk about it, tell you what it's used for, describe it, but can't put a label on it. Right-brained people go through life talking like this and the left-brained person is saying, "Just put a name to it so I'll understand what you're saying."

As a right-brained student, I often misused vocabulary words or interpreted them incorrectly. As I look back, some of the experiences are humorous. I was a junior taking English Lit. We were asked to give a report on our favorite poet. We had to describe the poet's background and explain why we especially liked this writer's poetry. I stood up and said, "I like this poet because he was so creative and marvelous and he had so many different styles. I could see that no matter what style he used, he handled it well.'' The name of the poet I chose was Anon. I was a junior in high school. I did not get laughed at, which was very special because it could have been a disaster. This is an example of the kind of things right-brained people do. I made all the right associations and said all the right things about style and meter and creativity — I just missed the name. To this day, when people ask me who wrote a book, I can't tell them — I can tell them the name of the book, or what's in the book, or what it looks like, but for some reason I can't remember the author's name.

Left-brained people usually have a good sense of time. They hate being late for anything. There appears to be a dichotomy between the right-brained person's sense of time and his ability to be on time. First, they appear to have no sense of time. They are usually late. They can go out for a newspaper and come back two days later. I often get so involved in what I am doing I forget I am supposed to be someplace or I think I have time to put one more load of washing in. I really try, but I usually end up late anyway. On the other hand, if I want to wake up at seven in the morning, I do not have to set an alarm. I have an intuitive inner clock — I just choose to ignore it occasionally.

Left-brained people think in the abstract — right-brained people often think analogically. I have raised two right-brained children and now I have a right-brained grandson. I discovered through trial and error that logic and reason do not work when disciplining these children. To say to them, "If you get all your work done, you'll be able to go out and play, and if you don't get it done, you'll go to your room," is usually a waste of time. They don't care if they go to their rooms or not. They would rather be there anyway, playing music, reading, or daydreaming. Your words don't register because it's not their problem at that point.

When my daughter was growing up, her room was always a mess! I threatened her. I took away the stuff she left out, but nothing worked. I gave her a puppy for her birthday. The puppy chewed up two pairs of designer jeans that were on the floor. That's all it took. It had become her problem! From then on, everything was picked up. Now, granted, the mess was all under the beanbag or in the closet, but at least I could not see it.

Right-brained people relate to an analogical approach to discipline and they react positively to humor. When I was teaching teenagers, I found that if I said to a student, "Finish your homework because . . . " it went right over his head. But if I said, "If you don't get your homework done, I'm going to unscrew your head and bounce it all the way to Siberia," the student would sit down and get to work.

Left-brained people go through life adjusting to life. They do what others expect them to. They tend to conform because functioning in a left-brained world is comfortable for them. Right-brained people try to change life to meet their basic needs, making them appear self-centered, which isn't true at all. They are considered rebels and non-conformists. Understanding that left and right-brained people approach life differently will help you become more aware of yourself. When my husband and I recognized our differences, we stopped trying to change each other. We learned to accept and enjoy each other. After all, together we make a whole brain.

*"Learning to love yourself
is the definition of change."*

Hugh Prather
Notes on Love and Courage

Destination

You Are Becoming
What You Are Going to Be

The process of learning to enjoy my husband's left-brainedness has helped me accept my right-brainedness. I no longer accept feelings of failure, hurt, and guilt for being different. I choose to celebrate being right-brained. I have discovered each person is unique and the center of his own universe. I know the world through my own experiences, not through someone else's. Changes and growth take place within this universe. Since it is my universe, I am in control of it. Therefore, I also have control of the changes and growth within my life. In his book, *Love*, Leo Buscaglia says, " . . . man is always capable of growth and change, and if you don't believe this, you are in the process of dying."

Join me. If you don't like what is happening in your life, change it. Start out by assuming the responsibility: you are responsible for your life and you are the only one with the power and responsibility to change it. You have to do it! No one else can do it for you!

Your mind is the seat of your intelligence. Your thoughts give your life reality. Your mind controls your emotions, attitudes, and actions. If you think something sad, you will feel sad. If you think you don't like something, you won't. If you believe you are a failure, you are.

Your mind controls your life, but YOU CONTROL YOUR MIND. Think of the saddest thing that has ever happened to you. Feel the sadness. Feel the hurt. Now, picture in your mind a beautiful snowflake or flower. Concentrate on it. What happened to the sadness and hurt? Did they leave?

If you want to feel happy, loved, or successful, think about the things that give you happiness, make you feel loved, or help you achieve success. Look for the good. You will find what you are looking for. The way you use your thoughts determines your experience.

Some years ago, I had been told I had 150 allergies and would have to be very careful with my diet. I was allergic to sugar, all additives, food colorings, preservatives, coffee, alcohol, and junk food in general. I was in the hospital for treatment complaining loudly about my allergies. A nurse walked into the room and stood there listening. Quietly she said, "God must love you very much. You are only allergic to those things that none of us should eat anyway." My mind suddenly went "tilt!" I no longer have allergies, only an awareness of what I should and should not eat to be healthy. If you want to be free of a problem or situation, look for the opportunity. Discover the learning experience. Grow with life, don't complain about it. The more of yourself you find, the more you will discover there is to find.

I am often frustrated with the never-ending challenge of self-awareness. A few years ago I went for a job interview. I really wanted the job. I bought an expensive suit, had my hair done, and researched the background of the company. I even practiced my responses to questions in front of a mirror. I did not get the job. Instead of focusing on my feelings of inadequacy and rejection, I looked at the interview as a growing experience. I acknowledged that if I was supposed to have had that job, I would have gotten it. Instead of pushing, I let go. Two weeks later, I got the job of my dreams!

By letting go, I am learning to be all of me that I am at any one moment. Maybe I'll be lucky enough to discover more of myself in the next few minutes, but right now this is all I've got. Being totally myself puts me in the right job, the right relationship, and the right place for me at the time.

From the time we enter school, we are taught that everything we do must look like what everyone else does. As we progress through high school and even into college, conformity is the key word. Yet every time we decide to do something, we set in motion changes in our lives. Our tomorrows depend on our choices today. So fight back! Think for yourself. Make your own choices. Don't let others make them for you. Your inner voice will give you the answers. Listen! It will tell you what needs to be done to accomplish your dreams.

Remember, you are the only one who has to spend twenty-four hours a day with you. Make them count. If you are living the moment, you don't have time to worry about your past failures or what you might not succeed at in the future. Put all your energy towards doing the best you can now — then let go of the results.

Be gentle with yourself. The other day I was feeling especially down about "blowing" a television interview. Sometimes walking on the beach helps me re-energize and center. While enjoying the sound of the surf, I saw a young man jogging towards me. Written on his T-shirt were the words, "Be patient — God's not finished with me yet." Immediately I recognized the interview not as a failure but as an opportunity to learn and grow. As I analyzed the interview, I was patient with myself. I recognized that I had been trying to please the host and had lost me. A week later I had a similar interview which went beautifully.

Be aware that you may be so used to failure you become afraid of success. Lacking awareness, we become comfortable with unhappiness. This pattern has demonstrated itself several times in my life. When I was a little girl, I received love and affection from my mother only when I was sick. As I grew up and entered into male-female relationships, I found I would get ill whenever a relationship began to deepen. It took many years of searching for self-awareness before I realized the connection.

Recently , my inability to handle success surfaced again. I had just published *Unicorns are Real* and was beginning to experience positive feedback. Each time someone walked up and said something nice to me about the book, I would choke up, my eyes would water, and I would stop breathing. What should have been a happy experience was painful. I am learning, step by step, to feel comfortable with happiness and success.

Your imagination is a gift, celebrate it! Albert Einstein said:

"Imagination is more important than knowledge. For knowledge is limited, whereas imagination embraces the entire world, stimulating progress, giving birth to evolution."

Embrace the world!

Acknowledge your intuition. Intuition is a special kind of knowing. It is your inner voice. Listen to it! When George Washington Carver was asked how he knew what to do with the peanut, he said, "I asked it." That's all. His discoveries helped millions. Your intuition is valid!

You will always be what you believe you are. The way out of the maze of failure is to believe in yourself. Know that you can be whatever you want to be. When you are aware of yourself and your unlimited potential, you can look at the world and know there are no limitations, only different ways of accomplishing the same thing.

I am an individual . . .
my own distinctive self
with my own identity
I am like a snowflake
that is different from all others,
with abilities and talents,
emotions and opinions,
ideals and beliefs . . .
all shaped into a pattern
never seen before.

Karen Raun
Our Inward Journey

"All I want to do, need to do, is stay in rhythm with myself."

Hugh Prather
Notes to Myself

Riding The Wind
A Right-Brained
Approach to Stress

Although I am growing and learning to celebrate my right brain, I still experience stress living in a left-brained world. I still am pulled between doing what is expected of me or going ahead and changing my world to meet my needs. You, too, may be having difficulty dealing with the left-brained world. Like me, you may be experiencing stress. Sometimes I am not organized when I feel a need to be. I do not do things in sequential order. My friends do not consider me especially logical. They often laugh at the way I think. I have difficulty allocating a certain amount of time to writing each day. My creativity comes in erratic patterns, often at three or four o'clock in the morning. I am always late. I not only march to a different drummer, I run!

I have read all the books I could find on organization and stress. None of them worked for me. How can you organize more efficiently if you aren't organized to begin with? How can you set priorities when you can't remember what it was you wanted to do that was so important? The more I read, the more uptight I became and the worse I felt.

I found the left-brained world full of all sorts of things that right-brained people don't do. Right-brained people constantly are asked to live up to expectations they are not comfortable meeting: sitting still, balancing a checkbook, wearing the same style clothes everyone else wears, remembering to pay the bills. Sometimes I finish projects in my head and lose the desire to actually complete them. I have found that the words "should" and "must" cause me a great deal of stress. Being right-brained, I do not want anyone telling me how to run my life. When I read books such as *Passages* that tell me what stages I go through in my life, my first reaction is, "Want to bet?" I then deliberately set out to prove the book wrong by making sure I function differently.

I remind myself of my two-year-old grandson. If I want him to do something, I tell him not to do it. The other day he was at my house for dinner. We were having peas, mashed potatoes, and chicken. He usually mashes half the peas into the potatoes and flips the remainder across the room. He is always careful not to eat any. This time I decided to tell him he was not allowed to eat even one pea. He ate over half!

I am not suggesting this as a form of controlling our children, but as an example that I often act two years old when I react to people who try to control my life.

I have found my reaction is different if I give them permission to control. I do not feel stressed. I might say, "I would like your opinion on how I should handle this situation. I'll make a better decision if I am aware of your point of view."

Stress is an individual thing. What upsets *me* may not bother *you* at all. A messy desk at work drives my secretary up the wall. I get upset if I can see the top of my desk — but I can always find everything. Although I'm usually late, being late upsets me. Other right-brained people could care less. They'll get there eventually. You cannot pick up a book and assume what it says is good for you. What the author has written about are the things that make him or her stressful. The first things you need to look at are those that cause *you* stress. If being late doesn't bother you, then don't worry about being on time. If it bothers you, do something about it.

As you do something about it, try not to fight your right-brainedness. If you are not an organizer yet, try to organize your life to please others because you think you should, you're going to induce more stress. Learn to take what you have, to understand the way you function, and go with it. Be willing to accept that functioning randomly is okay. Being unorganized is okay as long as you get things done.

A consultant called me the other day and said she was giving a weekend workshop for administrators. She said, "We'll help you get your life organized so you can get more done." I said, "I'm not interested." She said, "Why not — I've heard you have problems with organization." I said, "That may be true but I certainly don't want to get any more done than I am already doing."

54

Dr. Norman Vincent Peale and his wife defined stress as "your reaction to a situation or an idea. It can be either a positive reaction or a negative reaction. Negative reactions can drag you down, whereas positive reactions can give you a chance to grow."

Too much positive stress or too many positive things happening in your life can be as bad for you as too much negative stress. I haven't yet learned to unconditionally love myself. Accepting love from others is stressful for me.

Sometimes stress is not what it appears. When you feel physical pain, it's an indication that something's wrong. When you experience stress, it's also an indication that something's wrong. Every time someone said to me "I think you're great!" I cried. But I listened, and I realized that the problem was the way I felt about me. I learned the only way I could reduce the stress was to look inside to find the cause.

Sometimes we're unable to look inside. About five years ago I decided I was going to become more loveable. Over and over again, I told myself, "I am going to become more loveable. I am going to become more loveable." I gained thirty pounds. The most loveable person I ever knew was my grandmother and she was heavy. By saying I wanted to become more loveable, I was thinking I wanted to become more like her. This thought process was internal; I wasn't aware of it.

DISCOVERING THE CAUSE

If you are right-brained and are experiencing stress, try to understand that part of your stress is caused by not having the whole picture. You just have pieces of it. You've got to figure out some way of getting a picture of the whole situation before you can let go of that stress. One of the ways to see the whole picture is to sit down and think of whatever it is that is causing you stress. Take a piece of paper and start writing. Write down anything that comes into your head — words, whole ideas, or small phrases. Get every thought down on paper. Read what you have written. Is there a main problem? Are your thoughts falling into a pattern? Is the stress caused by lack of money, or by the fact that somebody doesn't know how to handle money, or both? Take the time to discover where you're going and where your stress is coming from.

RELEASING ENERGY

Let's start with some simple stress removers. Electrical energy builds up inside your body when you are in a stressful situation. Luckily, you have some energy release points. Those are points on your body where you can actually get rid of excess energy. Try to locate these points. One of them is on the back of your neck. By rubbing the back of your neck, you may immediately release energy. If someone you love and care for is uptight, try rubbing the back of his or her neck. You may notice that person beginning to relax.

Have you ever seen an uptight person rubbing his or her arm? There are energy release points on the insides of the elbow. Another release point is behind the knees. If you are in a stressful situation and you can't start rubbing, just swing your legs or get them in a position where they rub on the chair right behind the knees. You may find yourself beginning to relax. And, of course, there are the bottoms of your feet. Many people love to have their feet massaged or their backs rubbed. These are energy points that have been located through kirilian photography. These areas of the body have energy coming off them. Reduce your stress by learning to use these energy release points.

RELAXING YOUR EYES

During a normal day, your eyes may become tired, especially if you work on a computer, do paperwork or watch soap operas all day. To relax your eyes, close them. Put your head as far back as it will go. Don't put it so far back that it hurts. Keep your eyes closed. Look up at your forehead as far as you can look. Look down at your chin. Turn your eyes to your left and to your right. Move your eyes slowly. Now roll your eyes counter-clockwise. Reverse and go clockwise. Repeat the entire exercise as many times as is comfortable. Three times usually is sufficient. Bring your head forward very slowly. You'll find an immediate release of tension around the eyeballs and eyelids. Do you feel the difference?

CENTERING YOUR BODY

We each have a physiological center of our bodies. Everyone's center is different. The first thing to do is to find your own center. If you stand up, you can actually feel your center. Put your feet so they are directly under your shoulders and your body is in balance. Drop your hands to your sides and let them relax. See if you can feel your center; it is usually between the lower two ribs. Imagine the center of balance and put one hand on that spot. Now, I want you to imagine you are a very little person and you are all inside yourself. Just feel yourself intently and feel your emotions. Put all of yourself into that tiny spot. While you're in there, I want you to make yourself feel good. When you feel as good as you can possibly feel, let the goodness begin to grow until you have filled up your whole body and beyond. When you're all filled up, be aware of how good you feel. This is your beingness — the real you.

BREATHING

Breathing is another way of relaxing. I read some neat research the other day about breathing. With the invention of the CAT Scan came the ability to take pictures of the brain. Scientists have discovered from CAT Scans that when you breathe through your left nostril, you activate the right side of your brain. When you breathe through your right nostril, you activate the left side of your brain. Perhaps the yogis who practice controlled breathing are alternately activating both sides of the brain, creating a feeling of balance and wholeness!

Sequence breathing may be a way to activate the left hemisphere. In a left-hemispheric experience of breathing, you breathe in rhythm and count. Ready? Breathe in, 1, 2, 3, and up to the count of 6. Breathe out 1, 2, 3, 4, 5, 6. Breathe in, 1, 2, 3, 4, 5, 6. Breathe out, 1, 2, 3, 4, 5, 6. Are you beginning to relax or are you beginning to feel stressed?

Many people who try rhythmic breathing — especially if they're right-brained — find it makes them feel more uptight. They are driven up the wall when they are told to breathe in so many times and breathe out so many times. Free breathing may be a better way to activate the right hemisphere. Close your eyes. Breathe the way you feel like breathing. Take nice, deep breaths and let them out in whatever rhythm feels most comfortable — but don't count. Breathing deeply and naturally relaxes the heart muscles, increases the oxygen in the blood, and generally relaxes the body. Breathe whichever way feels best, but *do* breathe!

COLOR BREATHING

There are some other ways you can breathe. Right-hemispheric people relate to color. Try breathing in a rainbow. Start with the deepest color red, breathe red in and let it out. Let it take all the tension with it. Then breathe in orange and let it out, along with any negative thoughts. Breathe each color of the rainbow and when you get up to violet, you will be flying high. You will feel emotionally up and energized. This is one stress-reducing activity to use about two o'clock in the afternoon when you think you're going to fall asleep.

COLOR CENTERING

Color has a definite effect on people. Researchers are finding that people in jails react strangely to the color pink. When some offenders are arrested and become violent, they calm down immediately if they are put in a pink jail room. There is a slight problem. The effect appears to last only about twenty minutes.

The following activity will help you discover the effects of color. Put your feet right under your shoulders and put your left arm out. Look directly at a specific color. Have a friend place his left hand on your right shoulder and his right hand on your extended left arm near the wrist. Remain looking at the color. Try to resist as your friend attempts to push your left arm down. If your arm was strong, the color is a good one for you. If your arm was weak, I would not encourage you to surround yourself with that color. Certain colors take away your energy and emotionally bring you down or make you feel high. Those colors vary from person to person. Be aware of what color relaxes you and what color helps you work faster or be more creative.

VISUALIZING COLOR

Sometimes you can't get to your color. That's when you want to learn how to visualize the color. In your head, I want you to see your favorite color. Notice how your body feels. Now I want you to pick a color you absolutely dislike and visualize it. Can you feel what's happening to you? Now, go back to your favorite color. Can you feel the changes in your body? When you feel tension building, stop and visualize your favorite color.

SMILING

You can lower your stress simply by surrounding yourself with positive objects or by smiling. Experience the effects of positive and negative emotion on your strength. Do the following with a friend. Look at your friend's face. Close your eyes and have her test your strength by pushing down on one arm. Open your eyes. Have your friend frown at you and test your strength again. Did you feel yourself get weaker? Most people lose strength in their arm when frowned at. Now have your friend smile at you and test your strength again. You should feel stronger. Think about what you're doing to yourself when you frown. Consider what you're doing to the people around you when you frown. If you are strong when frowned at, you may be more comfortable with negative reinforcement than positive.

Medical science has discovered that people who smile most of the time experience little stress and few stress-related diseases. Doctors who work with cancer patients tell them to smile. It doesn't make any difference what's happening to you when you smile. Your body gives off certain endorphins that affect you. When you frown, enzymes are created, often high-acid ones affecting your stomach and other parts of your body.

One teacher I worked with tried an interesting experiment. She had her students make smiling faces in their favorite color and placed them on their desks. The test scores all went up. The students were told to look at the smiling face when they began to feel uptight or couldn't remember an answer. Color and positive images do amazing things to their stress factors as well.

SPORTS PSYCHOLOGY

There is a field of psychology called sports psychology. Two good books on the subject are *The New Golf Mind* and *Inner Game of Tennis*. A football coach attending one of my lectures asked what effect using red on uniforms would have. I don't know, but I have heard that some professional teams are using the visualization of color. They found that visualizing red made the players more aggressive while visualizing blue calmed and centered them. A weight lifter I know discovered he could lift fifteen pounds more than usual by visualizing red. The effects are different for each person. We're just touching the surface of what color and positive and negative images can do.

GUIDED RELAXATION

Have you ever bought a relaxation tape that says to start at your feet and relax? Go to your knees and relax? The approach is sequential and usually turns me off. When I teach relaxation, I'll say, close your eyes, go to the part of your body that tells you it needs relaxing and relax it. When that is relaxed, go to the next part of your body you feel needs relaxing. If you are right-brained, you will find working randomly, rather than from your feet up to your head, will enable you to relax more quickly. Get rid of the sequence. Get rid of the logic and do what you feel. Go to the next part of your body that needs relaxing. Maybe your whole body doesn't need relaxation, maybe only a little part needs it.

Using "links" also helps you relax. Use a favorite flower; a smell; a color; a favorite place or even a saying in a book. All of those things, if used consistently when you relax, will become "links" to relaxation. You will find when you experience or visualize your link, you will instantly relax no matter where you are. A link may be anything — the Lord's Prayer, a song, or simply a color — as long as you choose it and use it consistently.

MUSIC

Listening to music is relaxing to many people. Classical music written in four/four time and Stephen Halpern's music relax me. Halpern's music is unique. It is free-floating. Using research and an intuitive understanding of how sounds affect the human mind and body, he has composed and produced music that triggers a deep state of relaxation. When asked about music, Stephen will tell you, "For me, music has been the key which opens that part of me that is joyous, unbounded, and at one with the universe."

POTPOURRI

Some relaxation techniques belong in *Ripley's Believe It Or Not*. My suggestion is, try the following activities. If they work, use them. If you are in a stressful situation, try placing your tongue on the roof of your mouth. Some people immediately stop feeling the stress. Try it — see if you feel the difference. Your jaw has to drop and you can't clench your teeth and your whole face relaxes.

Take a bath in your favorite color. Use food coloring to color the water. Take a walk or engage in some form of physical exercise. Buy a pet. Medical science is finding that people with pets tend to live longer. Meditate or say a prayer. Whatever method you use, enjoy it.

*"If I have inside me the stuff
to make cocoons — maybe the stuff of
butterflies is there, too."*

"Yellow" in Trina Paulus'
Hope for the Flowers

No Clouds In The Sky

Right-Brained Strategies for Living in a Left-Brained World

Being right-brained is not a disease, it is a gift. Although our educational system and much of the business world see right-brained people as unable to cope with the system, they forget that we create what they teach or manufacture. We are the poets, the writers, the inventors, the artists. We are the changers of the world. History is written about our adventures. Writers find their material woven within our lives.

Yet, many of us spend our time complaining about the "left-brained" things we can't do. Forget what you can't do. Forget you can't spell. You can always hire a good speller. Concentrate on your strengths. If you can't think of any, create them. Enroll in an art class. Join a creative club. Keep a journal of your thoughts. You may find them quite poetic. Take tennis lessons. Begin running. Write a book. You might be as lucky as I am.

Realize that you can use those strengths to accomplish anything you choose. If you have a strong sense of color, color code everything. If you are hyper, use that movement to learn. Work with your own learning style, and open the door to unlimited potential.

I would like to share with you some of the ways I have found to become more aware of my potential. They are not meant to be the only ways. I am not a guru to give you answers. I only want to increase your awareness and show that there are many ways to do any task. All roads to the same house get you there.

RIGHT-BRAINED STRATEGIES FOR WORK

Most jobs require you to be productive from nine to five. I was always frustrating my boss and myself because I couldn't conform to these time constraints. My productive periods happen to be different. I am not a morning person. I don't wake up until ten o'clock. When I finally begin functioning, I do well until about one o'clock in the afternoon. I stopped taking lunch when everyone said I should because it wasted an hour of my peak productive time. From one o'clock on, I'm shot until the evening. From about eight to midnight, my motor is running again. Sometimes, it keeps running until three or four o'clock in the morning. I've learned when my peak periods are. I don't tackle a difficult task during a non-peak period. Instead, I think about the task, walk and think, or just roll ideas around in my head. I find non-peak periods are excellent times for lunch. Knowing I won't be accomplishing anything anyway, I can really relax.

Know your productive periods and work with them. Trying to conform to someone else's idea of time only leads to frustration and stress.

Right-brained people like things organized and neat but things still seem to fall apart if they even look at them. Recognizing your right-brained strengths and using them to create a form of structure helps you organize.

My office desk has always looked like the local garbage dump. I could usually find everything but every morning I would spend an hour leafing through the piles trying to decide what I needed to finish or start. Recognizing that I think in large clusters and relate to colors, I bought five colored folders: red, green, blue, yellow, and white. I fit every paper into one of the folders. The red folder means emergency — it should have been done two days ago. The green folder means complete today. The blue folder indicates pending. Everything in the white folder needs to be filed. Finally, I no longer need to deal with a messy desk. I come in and there's the red folder. I pick it up and I do it.

Oh, I forgot the yellow folder. The yellow folder holds what I really want to do. It's my reward for opening the red and green folders first. Sometimes when I can't get started, I allow myself fifteen to thirty minutes in the yellow folder. Doing this seems to create the right frame of mind for attacking the red folder.

I remember spending a whole day in my office filing everything in alphabetical order. I spent the next month looking for everything. I have since discovered that I cannot deal with anything in alphabetical or numerical order. To solve the problem, I set up my filing system similar to my desk-top system, using color and large categories such as "Correspondence," "Promotion," and "Consulting." The file folders within each category are colored. Within each large category are smaller categories such as "Workshops Scheduled," "Workshops Pending," etc. They are color-coded with tabs to match the color of the larger classification. I no longer have any trouble finding a file or re-filing it. Green folders go back in the green area, etc.

I forgot to mention that two secretaries have quit because they couldn't handle such a logically-organized system!

I am a creative speller. I have been known to find several creative ways to spell the same word within one letter or memo. I have resorted to using the *Poor Speller's Dictionary*, a small pocket-sized dictionary in which the incorrect spellings of words are listed alphabetically followed by the correct spelling and a definition. Finally, I can use a dictionary.

I was trained to spell words by their sounds. Since I have difficulty hearing and pronouncing words, the method didn't work. One day I met Johnny, who seemed to understand how right-brained people see words. He attacked a word by finding all the parts he already knew how to spell. That way, he only had to learn a small part of the word. Instead of becoming "con-sid-er-ate," the word "considerate" became "con-side-rate." Suddenly, I could spell.

If you are anything like me, you often finish reading an article or chapter in a book and haven't the slightest idea what you've read. Recently, I tried reading Jean Houston's book, *The Possible Human*, and ended up feeling like an impossible human because I couldn't understand it. The problem was not the book but me. A few days later, I picked up the book and accidentally started at the back of the book. (I read magazines this way.) I read the Epilogue and all of a sudden everything fit into place. I could understand the first part once I grasped where the book was heading.

I have problems writing memos and letters. I cannot sit down and start from point A and go to point B and express what I want to say without first knowing where I am going. When I was with the school system, I learned to alleviate this problem. I kept several folders in my desk drawer. In one were memos my boss had written. The second one was a folder with all the memos the deputy superintendent had ever sent out. The third one contained the superintendent's memos. When one of them asked me to write a memo, I pulled the folder with the memos from that person and spread them out. I took a paragraph from this one and a paragraph from that one and I wrote the most beautiful memos you ever saw. One of my bosses said to me, "I have never known anyone who writes memos as well as you do." I learned that I've got to have some kind of a visual model, a sample of an end result, before I can get started. I've got to have all the pieces before me first.

Although many right-brained people have lots of ideas floating around in their heads, they often have difficulty putting them into any kind of story or book. When I was struggling with the manuscript for *Unicorns Are Real*, a friend shared two helpful ideas. Keep a notebook with you at all times. When an idea hits you, whatever the topic, write it down. If you forgot your notebook, use a napkin, boarding pass, or the back of a deposit slip.

If you have a deep discussion with someone, tape it. I found my best ideas come when I am giving them to someone else. When the notes and tapes become unmanageable, have someone transcribe them. Next, collect several boxes. Label them: Chapter I, Chapter II, etc. or write a general idea on each.

Cut (or you may prefer to tear) the notes apart and throw them into the chapter you think they belong in. Finally, take all the scraps of paper in each box and put them into the order that feels right. Tape these together in the form of a cross and send them to be typed. I did this and my left-brained husband's left-brained secretary threatened to quit.

Being a visual learner and having a desire to learn means I read a lot. With my schedule, I realized I would either have to lengthen the day or increase my reading speed. After considerable thought, I decided it would be more practical to work on the speed. Working with children and adults, I found there were a number of techniques that seemed to increase both speed and comprehension. The effectiveness of each technique varied from person to person. Sometimes only one worked and sometimes different techniques worked on different days. Try all of the following and pick the technique or combination that works for you:

- use a blue plastic overlay to cover your reading material.
- read with a blue or green light bulb shining on the page
- place your right hand on your left ankle or vice-versa while reading
- place a piece of paper in your favorite color under the book
- try visualizing your favorite color while reading
- turn the book upside down and read
- hold the book vertically; read top to bottom or bottom to top
- read only the key words (usually the nouns and verbs); rephrase the meaning in your head

•read words at random on the page; see if you can get the meaning

•try lying on the floor or in a beanbag chair while reading

Tony Buzan has wonderful examples of right-brained and whole-brained thinking in his books, *Use Both Sides of Your Brain* and *Make the Most of Your Mind*. He uses the following exercises to help people increase and acknowledge their creativity.

Within two minutes, write down all the uses you can think of for a paper clip. The average number of uses for most people is eight. If you thought of sixteen, you are very creative. Allow your mind to consider the fact that the uses do not have to be logical, nor does the paper clip have to be the normal sized metal object we are familiar with. When you have done your mental "reset," try writing down a list of the ways you cannot use a paper clip. You will find there is a way to use one for almost anything.

At one of Tony's recent workshops, a lady said you could marry a paper clip. When challenged, she informed us that marriage was an emotional contract and that not only was it possible to be married to a paper clip, her ex-husband had done it!

RIGHT-BRAINED STRATEGIES FOR LEARNING

As a student, many times I would study for days only to sit down to a test and find my brain completely empty. I knew I knew the material, but it just wasn't there and nothing I did seemed to bring it back. I became frustrated with the fact that the information seemed to be going in while I was studying but it couldn't get out when I needed it. One day it occurred to me if I could put the information into as many areas of my brain as possible, I would have a better chance of recalling it later. I experimented and came up with the following method:

1. Read a small portion of the information. Connect it to something you already know.

2. Close your eyes and repeat it, rephrasing it in your own words.

3. Keep your eyes closed and write the information in the air or visualize yourself writing it on a chalkboard.

4. Continue to keep your eyes closed while you visualize the material as a whole.

5. Finally, with your eyes closed, write the information on paper.

6. Several times within the next twenty-four hours, try to visualize or re-state the information.

7. Move your eyes as if your face was a clock. Stop at each number and recall the information. Make a note of which eye movements are the most helpful in recalling the information. Also, try to match the different ways you recall — visually, auditorily, etc. — with specific eye movements.

I recently learned some test-taking tricks that help. First, before I even look at a test, I write down everything I can remember from studying on a piece of scratch paper. I write dates, key words or phrases, etc., and am amazed how much I can remember if I am not searching for "answers." When I begin the test, I simply look for the answers in what I have written.

I begin the test at the end and work towards the beginning. I have found the answers to many beginning questions of the test hidden in the questions at the end. If I come to a question I don't know, I skip it. This way, I don't get uptight. When I have answered all the questions I know, I do a mental relaxation or visualize my favorite color. I put down the first answers that come into my head. If none come, I guess.

Getting through college was difficult for me because I was unable to take adequate notes. As I began to work with children, I discovered many of them drew pictures when they should have been listening. I recognized the fact that I did the same thing. The idea hit me: why not teach the children, and myself, to draw pictures explaining what the teacher or speaker was saying. I found that by doing this I could remember what was said almost word for word.

Tony Buzan, in *Use Both Sides of Your Brain*, teaches a form of note taking called mind mapping. Mind mapping usually starts in the center of the page with the main idea or topic. Springing out from the center are different ideas explaining, clarifying, or expanding the basic idea. Each branch from the center is drawn in a different color. The map reminds me of a flower surrounded by several leaves. The veins of each leaf form the design to write key words on. The small veins can be used for supporting ideas or connecting words. Tony suggests the use of color, texture, arrows, shapes, and pictures to trigger mental connections. Mind mapping allows you to see the concept as a whole.

Gabriele Rico, in her book *Writing the Natural Way*, teaches clustering. Similar to mind mapping, it accesses the right brain and allows creativity to flow freely. Either method can be used for note taking, planning, brainstorming, or creative writing.

87

As a teacher working with children and adults with learning problems, I have discovered some techniques that improve oral reading. In one instance, I worked with a woman who was a poor oral reader. I had her stand up and start reading from a book written on a third grade reading level. She seemed to be reading at about a second grade level. As she read, I noticed her hand kept jerking, and so I asked her to move her hand in some type of sign language while she read. She did and instantly she began to read beautifully, with complete ease. Then I held her hand behind her back and wouldn't let go of it. Her reading reverted to the second grade level!

If you are having difficulties with oral reading, try doing some sign language with your hands as you read — you may experience a great improvement.

Although many right-brained people are good at math, others have developed math phobia. To get rid of math phobia, begin by realizing that there are many different ways to get the same answer. Erase the traditional methods from your mind and pretend you have the freedom to decide how specific math problems should be done. Write the numbers 6, 2, 8, 3, 4, 5, 7, 1, 9 and see how many different ways you can use to find their sum. Try pulling out 10's, the 8's, and multiplying, grouping, etc. Arrange the numbers from 1 to 9. Notice that by connecting opposite numbers, you can figure the answer by multiplying 4 times 10 and adding 5.

In *Make the Most of Your Mind*, Tony Buzan has a whole section on alternate approaches.

$$6\ 2\ 8\ 3\ 4\ 5\ 7\ 1\ 9$$

$$1\ 2\ 3\ 4\ 5\ 6\ 7\ 8\ 9$$

$$1+9 = 10$$

$$2+8 = 10$$

$$3+7 = 10$$

$$4+6 = 10$$

$$40 + 5 = 45$$

Right-brained people are haptic. That is, they move a lot. In school, they are always in trouble for not being able to sit still in class. As adults, they are thought of as "hyper" or "highstrung." Not only does the left-brained world not understand that many people need to move to learn, it seldom provides the opportunity to move in a learning situation. Children are placed at desks and told to sit still. College students are placed in chairs and expected not only to sit still but to absorb what they are hearing. The result often is failure.

I have worked with many haptic adults. Joan was a pre-med student who was experiencing considerable stress. She was on the verge of being asked to drop out of the program due to poor grades. Joan was a right-brained haptic learner. I recommended that she buy an exercise bicycle, put her book on the handlebars and pedal like crazy. She made the dean's list.

Lynn was a college student majoring in English. Although she loved exploring the many facets of literature, she did not do well on the tests. She, too, was failing. She was a right-brained learner who learned not only by moving but by listening. I suggested she tape her lessons and go jogging. Her grades rose to "A's" and "B's." Make your need to move work for you, not against you.

Many right-brained people sing, hum, or make strange noises when trying to accomplish a difficult task. Go one step further. Learn to use music to increase your effectiveness. When you are doing a report, studying, or working on any left-brain activity, turn on your radio or tape recorder. Experiment with rhythms and types of music. The effects vary from person to person.

If you are studying music, using color may help. I have always loved to sing but have never been able to sing in tune. I am unable to hear the differences in pitch. I analyzed the way I was listening to music and discovered I was seeing colors as I listened. I was able to identify a specific color for each note. When I visualized the particular color for a note, I found myself singing in perfect pitch.

Although this approach may not work for you, it may open doors to other ideas that will work. Whether or not you ever learn to sing in tune doesn't matter. If you enjoy singing, then sing. Perhaps *you* are the one in tune and everyone else is off key!

RIGHT-BRAINED STRATEGIES
FOR THE HOME

Daydreaming is one of my favorite pastimes. I can be any place I want at any time and experience it in detail. In order to daydream, I must visualize or imagine visual images within my mind. Since it has finally dawned on me that I can take anything I naturally do and turn it into a tool to accomplish more difficult tasks, I began to use visualization as a memory tool. I have found that visualizing as I read increases comprehension. I can memorize a list of words by combining them into a picture. For instance, I take the grocery list — ham, butter, bread, jam, tomatoes and soap — and picture it as a pig stuffed with bread, glazed with jam and butter, surrounded by tomatoes with a cake of soap in its mouth. In *Teaching For the Two-Sided Mind*, Linda Williams has excellent suggestions for using visualization and fantasy as ways to record and remember information.

The frustration of grocery shopping is more than either brain can bear. Although I write them down, the things I need are listed in the order I run out of them. As I move up and down the store aisles, I must keep checking the list to make sure I haven't forgotten anything. Usually, I end up in dairy products realizing I forgot the shampoo.

Although I used to feel that my chances of surviving this world with such an "organization deficit" were slim, I now realize I can make others' organizational systems work for me. Someone put thought into organizing the grocery store and I decided to use their plan. I went shopping and copied the signs above each aisle in the order they appeared as I moved through the store. This was important since I move through the store backwards. (I don't want to squash the fruit.) When I completed the list, I had one hundred copies printed. I keep one on the refrigerator and check items off as I run out. I've cut my shopping time in half!

The kitchen is my least favorite room unless someone else is cleaning it. Not only do I not like to clean it, I never clean it twice the same way. Finding the measuring spoons is like hunting for a needle in a haystack. The disorganization of the kitchen so traumatized my husband, I had to find a workable solution.

First, I attacked the silverware drawer. I hate to stack and sort silver. I decided I could handle the sorting if I didn't have to stack it. I threw out the silverware divider and filled the drawers with large containers of various colors. Now, I simply dump all the knives in the container I have chosen to be their color.

Next, I attacked the cupboards. I tried putting the plates in once place, the cups in another, etc. Although this might work for you, it bothers my sense of artistic form. I need to see a complete set of dishes in one place. When I had completed my organizational plan, I found I couldn't remember which cupboards I had put which dishes in. Back to the drawing board. I needed color. I finally found I could buy doorknobs of different colors. Remembering that the pink rosebud dishes are in the cupboard with the yellow knob is easy!

By the way, the pans are still randomly thrown in the bottom cupboard.

Following a recipe for me is not only sheer agony, but downright near impossible. I either forget an ingredient or add one twice. Either way, my attempts at cooking never turn out well. One day, I decided to pretend I was a famous chef creating my own recipes. I let my imagination and creativity run wild. Not only did I have fun, the meal was delicious. I am not always successful at improvising, but neither are the chefs. When something does turn out well, I write it down quickly before I forget. For some reason, I can follow my own recipes. I have become known for my "Potpourri Soup" and "Leprechaun Cake."

My husband and I had been married about six months when he discovered my checkbook. In absolute disbelief he asked, "When was the last time you balanced this?" I calmly answered, "When did I open the account?" Needless to say, we have separate accounts. Why should I balance the checkbook when I know intuitively how much is in the account? I have never bounced a check on purpose and very few not on purpose.

I have discovered if I have a checkbook with check carbons or stubs attached to the check, I manage to write down the amount. I have also found that subtraction becomes easier by rounding the check amounts to the nearest dollar. Instead of subtracting a check of $12.34 from a balance of $924.70, I subtract $12.00 from $925.00. This method allows me to keep an approximate balance and a happy husband. There are now hand-held calculators that record your checks and deposits and maintain your balance. There also are calculators that play musical notes when you touch the keys.

I travel a good deal. Each time I prepare for a trip, I seem to have great difficulty deciding what to take. Frustration sets in and I take it all. I solved the problem in two ways. First, each time I decide on a wardrobe, I write it down. This way I can pull the same outfits at a later date without going through the decision-making trauma. If I take an outfit twice and don't wear it, I remove it from the list. Second, I keep duplicates of my make-up, hair dryer, hot curler, vitamins, bath necessities, and travel iron. Oh, yes, and most important, a traveling alarm clock! These never get unpacked. They always are ready to go. I also have six basic belts, six beautiful scarfs, and basic jewelry such as gold and silver chains and bracelets to match the belts. These, along with a beautiful robe and gown, also stay packed. I now can pack in thirty minutes flat!

"Actualization is having the courage to be the Gods within us."

Bob Samples
The Metaphoric Mind

Fly The Universe
Choose to Be Whole-Brained

Life is a journey, not a destination. I've journeyed from feeling no-brained to CELEBRATING my right brain. Sometimes I feel like a baby eagle about to take its first flight. Sometimes I'm still scared. Then I remember I am somebody special. The universe doesn't make mistakes. When I feel special, I want to fly high, to choose my destination, to laugh, to play, to be!

I have spent so much of my life trying to be left-brained that I got a pretty good picture of what a left-brained person should be. Now that I have acknowledged my right brain, I realize it, too, is limiting.

One morning I stood on a hotel balcony overlooking the bay in Corpus Christi. All I could do was give thanks for the change that had happened in my life.

I stood watching the sun slowly rise above the horizon, its rays vibrating with energy and illuminating the sky. I suddenly knew I could choose to be whole-brained. It was at that moment I *really* began to grow and change.

We all have the potential to express ourselves intuitively and logically. The two sides of the brain are joined by a natural bridge. They were designed to work together. When we learn to look at our experiences from different points of view, we begin to expand our consciousness. When we accept the possibility that we can move freely from one hemisphere to the other, from one mode of consciousness to the other, we are taking the first step to becoming total human beings.

Ornstein, the foremost brain psychologist, says, "When both brains are stimulated and encouraged to work in cooperation, the end result is an increase in abilities and effectiveness." The best doctor is the doctor who is highly trained in the latest treatments and medical research yet who uses intuition when diagnosing a difficult case. Great performers use their left brain to learn and speak their lines, but their right brain enables them to become the character they are portraying. The composers who hear songs in their heads must learn the structure of music to write them down so others can enjoy them.

To experience this process, imagine the energy flowing back and forth between your hemispheres. Visualize a color in your right hemisphere, its name in your left hemisphere. Try switching the two. Pretend you can activate parts of your brain at will. Use different sounds and colors. Can you visualize a color better at the front, back, or top of your head? Does sound work better to activate certain areas? Experiment — believe it is working!

Each of us has an infinite number of possibilities of connections within our whole brain. These connections form unique patterns of thinking, creating unlimited potential within us.

As I reached for my unlimited potential, I found I could think in musical notes. I experienced each person as a melody or a song. Every experience became a symphony. One afternoon, as I was listening to Stephen Halpern's "Dawn," color erupted into my consciousness. I found myself thinking in color. All aspects of my life took on color. When I thought of my friends, I saw colors swirling around them. I could identify a dominant color for each. I knew I could think in pictures and colors — could I also think in words, lines, shapes, or mathematical symbols?

Moshe Feldenkrais believed that freedom is achieved by the integration of mind and body. After spending a fascinating week with him, I began to consider my body as part of my intelligence. As I awakened my body, I began to think with it. My whole body was a thinking machine. I was aware that I not only saw and heard other people, I smelled, tasted, and physically reacted to them. Even my chemistry changed as, intuitively, I sensed a person. Animals sense and smell fear. Why can't we?

In the book, *Mister God, This is Anna*, Anna says the only difference between God and people is that God views things from all points while people have one point of view. When we begin to look at our universe from all points with both hemispheres, we begin to approach actualization. As I began to grow, I imagined people and experiences as holographs. When I faced a decision or difficult situation, I would walk around it, viewing it from all sides. You might try this, too. Fly above your problem and look down. View it as a painting or a play with different characters. Move under it. Look at its beginnings or roots. Discover its cause. I found Jean Houston's book, *The Possible Human*, helpful in learning new ways to view my universe. As I practice these exercises, my consciousness is expanding.

Everything in my life is becoming clearer and more beautiful as I experience it through both my brains. I am seeing the whole and the parts as a continuous circle. I am seeing these circles as parts of even greater circles. The grass is greener, the sun brighter, people are more beautiful. I am alive — the universe is mine to explore. I am flying free!

Free Flight

You who have traveled
and I ready to begin
guide to an ancient rumbling
I open my heart
and let the words come
a circle of knowledge is thus begun
let the words flow
let them turn like the sea
the blue and the silverfish
riding destiny
catches of glimmer
glimpses of shine
the light of this journey
this journey thru time
where life is illusion
and the wise only know
how the river turns
how the winds blow
and mirrors are mazes
and shadows are dreams
there's life in the waters
thru the reeds thru the reams

lift up from the water
free flight to the sky
vision of wings clapping
the eagle will fly
the dream and the dreamer
the whole and the torn
born thru the circle
I journey alone
this journey of knowing
once new and once known
we're born for the wanting
release from the stone
one spirit in motion
a wing on the waves
a shelter in the harbour
a harbour of caves
the dreamer, the believer
the holder, the rhyme
in step with the universe
a step out of time.

Mary Starr Whitney
16 November 84

Bibliography

Alexander, Thea, *2150 A.D.,* Macro Books, Tempe, AZ, 1976.

Andersen, Marianne S. and Savary, Louis M., *PASSAGES: A Guide for Pilgrims of the Mind,* Harper & Row, NY, 1972.

Andersen, U.S., *Three Magic Words,* Thomas Nelson & Sons, NY, 1954.

Ardell, Dr. Donald, *14 Days to a Wellness Lifestyle,* Whatever Publishing, Inc., Mill Valley, CA, 1982.

Asimov, Isaac, *The Human Brain, Its Capabilities and Functions,* Mentor Books, New York, NY, 1965.

Axline, Virginia, *Dibs in Search of Self,* Ballantine Books, NY, 1964.

Ayres, A.J., *Sensory Integration and Learning,* Western Psych., Los Angeles, CA, 1972.

Bach, Richard, *Illusions, The Adventures of a Reluctant Messiah,* Delacorte Press/Eleanor Friede, 1977.

Bailey, Alice A., *Education in the New Age,* Lucis Publishing Company, NY, 1954.

Bandler, Richard, *Frogs Into Princes,* Real People Press, Moab, UT, 1979.

Behrend, Genevieve, *Your Invisible Power,* DeVorss & Co., Marina Del Rey, CA, 1951.

Biffle, Christopher, *The Castle of the Pearl,* Harper & Row, NY, 1983.

Birren, Faber, *Color Psychology and Color Therapy,* The Citadel Press, Secaucus, NJ, 1950.

Birren, Faber, *Color and Human Response,* Van Nostrand Reinhold Co., Inc., NY, 1978.

Blakemore, Colin, *Mechanics of the Mind,* Harvard University Press, Cambridge, MA, 1977.

Blakeslee, Thomas R., *The Right Brain,* Anchor Press/Doubleday, Garden City, NY, 1980.

Blanchard, Kenneth and Johnson, Spencer, *The One Minute Manager,* Berkley Books, NY, 1981.

Bruner, Jerome S., *Beyond the Information Given,* W.W. Norton & Co., NY, 1973.

Bruner, Jerome S., *On Knowing: Essays for the Left Hand,* Atheneum, NY, 1973.

Bry, Adelaide, *Visualization,* Harper & Row, NY, 1978.

Buscaglia, Leo F., *Love,* Ballantine Books, NY, 1972.

Buscaglia, Leo F., *Personhood,* Ballantine Books, NY, 1978.

Buzan, Tony, *Make the Most of Your Mind,* Linden Press, Simon and Schuster, NY, 1984.

Buzan, Tony, *Use Both Sides of Your Brain,* Dutton, NY, 1974.

Buzan, Tony and Dixon, Terence, *The Evolving Brain,* Holt, Rinehart & Winston, NY, 1978.

Campbell, Don G., *Introduction to the Musical Brain,* Magnamusic-Baton, Inc., Saint Louis, MO, 1983.

Casebeer, Beverly, *Using the Right/Left Brain,* Academic Therapy Publications, Novato, CA, 1981.

Clark, Glenn, *The Man Who Talks With the Flowers,* Macalester Park Publishing Co., St. Paul, MN, 1939.

Course in Miracles, Foundation for Inner Peace, Tiburon, CA, 1975.

Dardik, Irving and Waitley, Dennis, *Quantum Fitness: Breakthrough to Excellence,* Pocket Books, NY, 1984.

Dass, Ram, *The Only Dance There Is,* Anchor Press/Doubleday, Garden City, NY, 1974.

DeMille, Richard, *Put Your Mother On the Ceiling,* Penguin Books, NY, 1967.

Dennison, Dr. Paul E., *Switching On,* Edu-Kinesthetics, Inc., Glendale, CA, 1981.

Diagram Group, The, *The Brain: A User's Manual,* G.P. Putnam's Sons, NY, 1982.

Dimond, S.J., and Beaumont, J.C., *Hemisphere Function in the Human Brain,* Wiley, NY, 1974.

Don, Frank, *Color Your World,* Destiny Books, NY, 1977.

Dyer, Wayne, *Gifts from Eykis,* Pocket Books, NY, 1983.

Edwards, B. *Drawing on the Right Side of the Brain. A Course in Enhancing Creativity and Artistic Confidence.* J.P. Tarcher, Inc., Los Angeles, CA, 1979.

Feldenkrais, Moshe, *Awareness Through Movement,* Harper & Row, NY, 1972.

Ferguson, Marilyn, *The Aquarian Conspiracy,* J.P. Tarcher, Inc., Los Angeles, CA, 1980.

Ferguson, Marilyn, *The Brain Revolution,* Taplinger Publishing Co., NY, 1973.

Flynn, *Mister God This is Anna,* Ballantine Books, NY, 1974.

Fox, Patricia L., "Reading as a Whole Brain Function," *The Reading Teacher,* October, 1979.

Gaddes, William H., *Learning Disabilities and Brain Function,* Springer-Verlag, NY, 1980.

Galwey, Timothy, *Inner Game of Tennis,* Bantam, NY, 1979.

Gardner, Howard, *Frames of Mind, The Theory of Multiple Intelligences,* Basic Books, Inc., NY, 1983.

Gawain, Shakti, *The Creative Visualization Workbook,* Whatever Publishing, Inc., Mill Valley, CA, 1982.

Gazzaniga, Michael S., *The Bisected Brain*, Appleton-Century-Crofts, NY, 1970.

Gazzaniga, Michael S. and LeDoux, Joseph E., *The Integrated Mind,* Plenum Press, NY, 1978.

Gendlin, Eugene T., *Focusing,* Bantam Books, Inc., NY, 1981.

Geschwind, Norman, "Language and the Brain", *Scientific American,* April, 1972.

Gibran, Kahlil, *The Garden of the Prophet,* Alfred A. Knopf, NY, 1978.

Gittner, Louis, *There is a Rainbow,* The Louis Foundation, East Sound, WA, 1981.

Golas, Thaddeus, *The Lazy Man's Guide to Enlightenment,* Bantam Books, Inc., NY, 1980.

Grady, Michael P. and Luecke, Emily A., "Education and the Brain,", *Phi Delta Kappan,* 1978.

Gregory, R.L., *Eye and Brain,* McGraw Hill, NY, 1974.

Gunther, Bernard, *Energy Ecstasy,* The Guild of Tutors Press, Los Angeles, CA, 1978.

Halpern, Stephen, *Tuning the Human Instrument,* Spectrum Research Institute, Belmont, CA, 1978.

Hendricks, G. and Wills, R., *The Centering Book,* Prentice-Hall, Englewood Cliffs, NJ, 1975.

Higbee, K.L., *Your Memory: How It Works and How To Improve It,* Prentice-Hall, Englewood Cliffs, NJ, 1977.

Hoff, Benjamin, *The Tao of Pooh,* A Penguin Book, NY, 1982.

Houston, Jean, *The Possible Human,* J.P. Tarcher, Inc., Los Angeles, CA, 1982.

Hunt, Roland, *The Seven Keys to Color Healing,* Harper & Row, San Francisco, CA, 1971.

Jampolsky, Gerald G., *Teach Only Love,* Bantam Books, NY, 1983.

Jampolsky, Gerald G., *Love is Letting Go of Fear,* Bantam Books, NY, 1970.

Johnson, Spencer, *The Precious Present,* Doubleday & Co., Garden City, NY, 1984.

Kaufman, Barry Neil, *Son Rise,* Warner Books, NY, 1976.

Leonard, George B., *Education and Ecstasy,* Dell, NY, 1968.

Leonard, George, *The Ultimate Athletic,* Viking, NY, 1975.

Lozanov, Georgi, *Suggestology and Outlines of Suggestopedy,* Gordon and Breach, NY, 1978.

Lupin, Mimi, *Peace, Harmony, Awareness,* Teaching Resources, MA, 1977.

Maclean, Paul D., *A Triune Concept of the Brain and Behaviour,* University of Toronto Press, Toronto, 1973.

Moss, Richard, *The I That Is We,* Celestial Arts, Millbrae, CA, 1981.

Mullen, Franklin and Chaffee, John, *Checkpoint 83,* Urban Ed 2000, Eastwood Printing & Publishing, 1983.

Muller, Robert, *New Genesis, Shaping a Global Spirituality,* Image Books, Garden City, NY, 1984.

Nelson, Portia, *There's a Hole in My Sidewalk,* Popular Library, NY, 1977.

Ornstein, Robert E., *The Psychology of Consciousness,* W.H. Freeman & Co., San Francisco, CA, 1972.

Ornstein, Robert E., *The Nature of Human Consciousness,* W.H. Freeman & Co., San Francisco, CA, 1973.

Ornstein, R. E., *Mind Field,* Grossman, NY, 1976.

Ornstein, Robert E. and Thompson, Richard F., *The Amazing Brain,* Houghton Mifflin Company, Boston, MA, 1984.

Ornstein, Robert E., Lee, Philip R., Galin, David, Deikman, Arthur, Tart, Charles T., *Symposium on Consciousness,* Penguin Books, NY, 1976.

Ostrander, S. and Schroeder, L., *Superlearning,* Delacorte Press and The Confucian Press, NY, 1979.

Ott, John N., *My Ivory Cellar,* Twentieth Century Press, Chicago, IL, 1958.

Paivio, Allen, *Imagery and Verbal Processes,* Holt, Reinhart and Winston, NY, 1971.

Paulus, Trina, *Hope for the Flowers,* Paulist Press, NY, 1972.

Piaget, Jean, *To Understand Is to Invent,* Grossman, NY, 1973.

Piaget, J. and Inhelder, B., *Memory and Intelligence,* Basic Books, NY, 1973.

Powell, John, *The Secret of Staying in Love,* Argus Communications, Allen, TX, 1974.

Prather, Hugh, *A Book of Games,* Doubleday & Company, Inc., Garden City, NY, 1981.

Prather, Hugh, *There is a Place Where You are Not Alone,* Doubleday & Company, Inc., Garden City, NY, 1980.

Prather, Hugh, *Notes on Love and Courage,* Doubleday & Company, Inc., Garden City, NY, 1977.

Prather, Hugh, *Notes to Myself, My Struggle to Become a Person,* Real People Press, Moab, UT, 1970.

Prather, Hugh, *The Quiet Answer,* Doubleday & Company, Inc., Garden City, NY, 1982.

Prines, Maya, *The Brain Changers,* Signet, NY, 1973.

deQuiros, Julio B., *Neuropsychological Fundamentals in Learning Disabilities,* Academic Therapy Publications, Novato, CA, 1978.

Raun, Karen, *Our Inward Journey,* Hallmark Cards, Inc., 1979.

Restak, R.M., *The Brain: The Last Frontier,* Doubleday & Company, Inc., Garden City, NY, 1979.

Rico, Gabriele, *Writing the Natural Way,* J.P. Tarcher, Inc., Los Angeles, CA, 1983.

Rose, Steven, *The Conscious Brain,* Vintage Books, NY, 1976.

Russell, Peter, *The Global Brain,* J.P. Tarcher, Inc., Los Angeles, CA, 1983.

Sagan, Carl, *Broca's Brain,* Ballantine Books, 1979.

Sagan, Carl, *The Dragons of Eden,* Random House, NY, 1977.

Samples, Bob, *The Metaphoric Mind,* Addison-Wesley Publishing Company, Inc., MA, 1976.

Samples, Bob, *Mind of Our Mother,* Addison-Wesley Publishing Company, Inc., MA, 1981.

Scott, Dru, *How to Put More Time in Your Life,* Signet, NY, 1980.

Segalowitz, S. and Gruber, F., *Language Development and Neurological Theory,* Academic Press, NY, 1977.

Silverstein, Alvin and Virginia B., *The Left-Handers World,* Follett Publishing Co., Chicago, IL, 1977.

Simon, Sidney, *Caring, Feelings, Touching,* Argus Communications, Niles, IL, 1976.

Skolimowski, Henry K., *The Theatre of the Mind,* The Theosophical Publishing House, London, England, 1984.

111

Smith, Adam, *Powers of Mind,* Random House, NY, 1975.

Sokolov, An., *Inner Speech and Thought,* Plenum Press, NY, 1972.

Springer, Sally P. and Deutsch, Georg, *Left Brain, Right Brain,* W.H. Freeman & Co., San Francisco, CA, 1981.

Tame, David, *The Secret Power of Music,* Destiny Books, NY, 1984.

Troward, Thomas, *The Creative Process in the Individual,* Dodd Mead & Co., NY, 1915.

Tulku, Tarthang, *Time, Space and Knowledge,* Dharma Publishing, Emeryville, CA, 1977.

Turner, Charles Hampden, *Maps of the Mind,* Macmillan Publishing Co., Inc., NY, 1981.

Vitale, Barbara, *Unicorns Are Real,* Jalmar Press, Rolling Hills Estates, CA, 1982.

Vygotsky, L.S., *Thought and Language,* The M.I.T. Press, MA, 1962.

Welbeck, Karen and Jones, Alex, *Creative Thought Remedies,* Alex Jones, Ontario, 1980.

Wilbur, Ken(ed.), *The Holographic Paradigm and Other Paradoxes,* Shambhala Publications, Inc., Boulder, CO, 1982.

Williams, Linda Verlee, *Teaching for the Two-Sided Mind,* Prentice-Hall, Inc., Englewood CLiffs, NJ, 1983.

Wilson, Sue, *I Can Do It! I Can Do It!,* Quail Street Publishing Co., Newport Beach, CA, 1976.

Wiren, Gary, *The New Golf Mind,* Simon & Schuster, NY, 1978.

Wittrock, M.C. and others, *The Human Brain,* Prentice-Hall, Inc., NJ, 1977.

Wittrock, M.C., Education and the Cognitive Processes of the Brain. In J.S. Chall & A.F. Mirsky (Eds.), *Education and the Brain,* University of Chicago Press, Chicago, IL, 1978.

Wittrock, M.C., Learning and the Brain. In M.C. Wittrock (Ed.) *The Brain and Psychology,* Academic Press, NY, 1980.

Wolf, Fred, *Taking the Quantum Leap,* Harper & Row, Inc., San Francisco, CA, 1981.

Wonder, Jacquelyn and Donovan, Priscilla, *Whole Brain Thinking,* William Morrow & Co., Inc., NY, 1984.

Young, J.Z., *Programs of the Brain,* Oxford University Press, NY, 1978.

Zukav, Gary, *The Dancing Wu Li Masters, An Overview of the New Physics,* Bantam Books, Inc., NY, 1979.

112

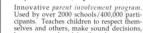

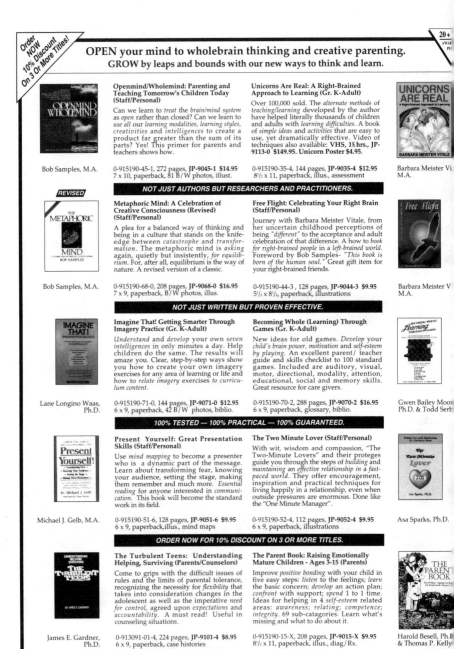